Let's Count Easter!

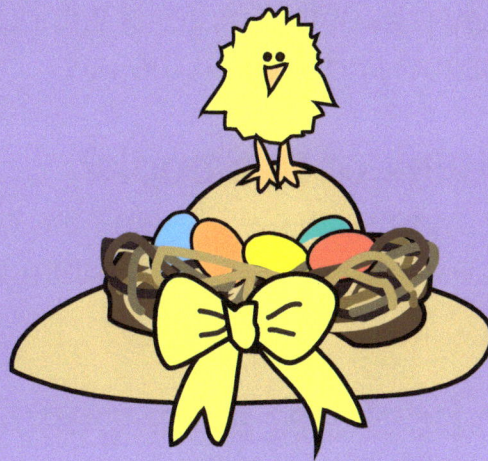

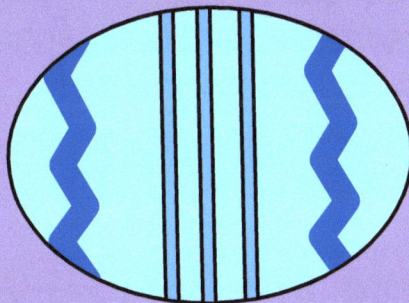

A Fun Kids' Counting Book
For Children Age 2 to 5

Alina Niemi

What do you think of this book?

Please consider leaving a review. It helps me get better and helps others decide if this book will help them. Thank you!

Find more fun books at alinaspencil.com

Plus links to stores with customizable shirts, stickers, mugs, bumper stickers, personalized pet bowls, phone and laptop cases, custom photo wall clocks, magnets, greeting cards, and much more!

ISBN: 978-1-937371-06-7

Alina's Pencil Publishing

1 one Easter basket

2 two nests

3 three bonnets

4 four children

5 five dresses

6 six lilies

7 seven dyes

8 eight lambs

maa
maa

9 nine
Easter eggs

10 ten ducklings

11 eleven flowers

12 twelve stripes

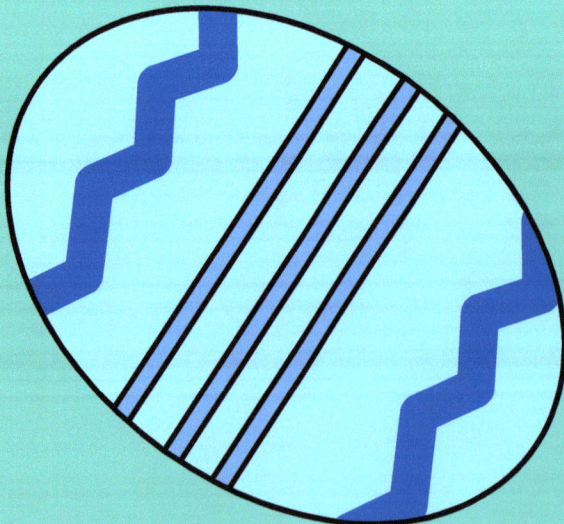

13 thirteen chocolate bunnies

14 fourteen chicks

15 fifteen rabbits

16 sixteen tulips

17 seventeen
hot cross buns

18 eighteen carrots

19 nineteen
blades of grass

20 twenty jelly beans

Check out the other books in the series!

Let's Count Trucks

Let's Count Trucks!

A Fun Kids' Counting Book For Toddler Boys And Children Age 2 to 5

Alina Niemi

Your truck-loving toddler boy or preschooler will have fun and learn to count with this educational book.

Find colorful pictures of pickup trucks, firetrucks, tractors, and more. Free coloring and doodling pages are also included in the print book.

Your child or grandchild can point to each object and count out loud. Many toddlers know their numbers but get confused when counting objects. This is a fun way to practice numbers 1 to 10.

Recommended for children age 2 to 5

Let's Count Halloween

Let's Count Halloween!

A Fun Kids' Counting Book For Children Age 2 to 5

Alina Niemi

Find favorite items you'll see during the Halloween holiday season, such as dancing skeletons, cute ghosts, and black cats. Kids will love the cute pictures and can practice counting numbers 1 to 20.

Recommended for children age 2 to 5

Also available:
Strange Ice Cream

Like ice cream but can't have dairy products? **The New Scoop: Recipes for Dairy-Free, Vegan Ice Cream in Unusual Flavors (Plus Some Old Favorites)** contains recipes for ice cream, sherbet, sorbet, and frozen yogurt, all without milk or eggs.

Try Peanut Butter and Jelly Ice Cream, Cucumber Mint Frozen Yogurt, or Pineapple Sherbet. Tropical flavors include guava, lilikoi (passionfruit), lychee, and mango. Learn to make mochi ice cream and yogurt at home!

Like to draw?

Learn about Hawaii's culture, food, and animals while you doodle. There are over 100 pages for you to color, draw, or decorate.

Contains how-to-draw tips, and a glossary, plus a guide to Hawaiian language and pronunciation.

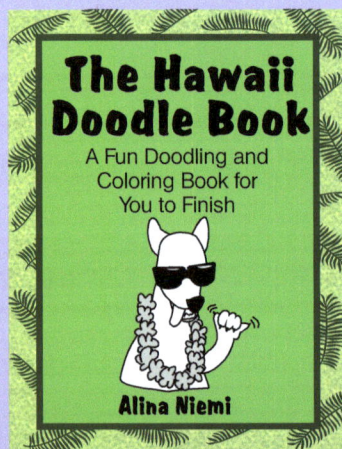

Available 2014:

When Masako's musubi in her lunch turns into a sumo wrestler, she tries to show her friend. But there is only a rice ball. Did she imagine it? Find out in **Masako's Bento Lunchbox Surprise**.

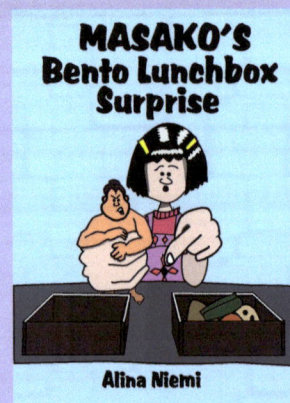

www.alinaspencil.com

22

www.ingramcontent.com/pod-product-compliance
Lightning Source LLC
Chambersburg PA
CBHW042113040426
42448CB00002B/251